T0093594

CORRUPT CULTURES
Cheating in Science and Society

CORRUPT CULTURES
Cheating in Science and Society

Roy Yorke Calne

World Scientific

NEW JERSEY · LONDON · SINGAPORE · BEIJING · SHANGHAI · HONG KONG · TAIPEI · CHENNAI · TOKYO

Published by

World Scientific Publishing Europe Ltd.

57 Shelton Street, Covent Garden, London WC2H 9HE

Head office: 5 Toh Tuck Link, Singapore 596224

USA office: 27 Warren Street, Suite 401-402, Hackensack, NJ 07601

Library of Congress Cataloging-in-Publication Data
Names: Calne, Roy Yorke, author.
Title: Corrupt cultures : cheating in science and society / Roy Y. Calne.
Description: [New Jersey] : World Scientific, [2022]
Identifiers: LCCN 2018037041| ISBN 9781786345608 (hardcover : alk. paper) |
    ISBN 1786345609 (hardcover : alk. paper)
Subjects: LCSH: Fraud in science. | Science--Moral and ethical aspects.
Classification: LCC Q175.37 .C35 2022 | DDC 174/.95072--dc23
LC record available at https://lccn.loc.gov/2018037041

ISBN 978-1-78634-561-5 (ebook for institutions)
ISBN 978-1-78634-562-2 (ebook for individuals)

British Library Cataloguing-in-Publication Data
A catalogue record for this book is available from the British Library.

For any available supplementary material, please visit
https://www.worldscientific.com/worldscibooks/10.1142/Q0166#t=suppl

I dedicate this book to my wife, Patsy and my brother, Donald

Acknowledgment

I am extremely grateful to David Cooper for writing the foreword, to Philip Ball for his skillful assistance with the images and to Nilu Karun for the photographs of the sculptures. I thank the following for their generous help with the text and their constructive comments:

Martin Birchall, Nadezda Babenko, Michael Kaabak, Susan & Donald Calne, Suzie Calne, Graham Howes, Allan Kirk, KO Lee, Stephen Lock, Allan Macdonald, Ben Mannings, Randal Morris, Paul Nicholson, David Parry-Smith, Mel Thompson, Maria Tippett, Tadashi Tokieda, William Wall.

I thank my Personal Assistants, Anne Parry-Smith and Jenny Richards, for their cheerful and extremely hard work on the continually changing text.

Foreword

David K.C. Cooper MD, PhD, FRCS, FACS

It is a great privilege for me to contribute these few words to this thought-provoking book written by one of my early surgical mentors who is undoubtedly one of the world's great pioneers in organ transplantation. His vast experience of medical research in both the experimental laboratory and in clinical practice makes his opinions on scientific fraud of very considerable interest.

From time-to-time, we all tell "white lies", often with good intentions. In this book, Sir Roy provides some hilarious examples. These are generally considered excusable, but less innocuous deception is not excusable. It is this topic, particularly as it relates to science, that the author addresses. He gives examples of fraud in medical research, but also in several other fields, such as art and wine. As he is an accomplished artist himself, with a wide interest in the field, his comments on artistic fraud are very relevant. However, fraud is of special and fundamental importance in science, as the aim of science is to identify "truths".

In this book he proposes that there are three types of truth:

1. **Manifest truths**: such as that we are all mortal, the sun rises in the morning and sets in the evening etc. On these truths there can be no rational dispute.
2. **Unverifiable truths**: these are based on myths, superstitions and wishful thinking. The manipulation of these types of truth in many different ways has been the pattern of establishing the organization of all nations and states, so that people can work together without mayhem.
3. **Scientific truths**: these are to be provisionally accepted or rejected according to evidence from further data. Science is a relatively new form of establishing knowledge as the word "science" means. Its novel attribute depends on independent reproducibility. The scientific method makes this type of truth quite different to the second category of unverifiable truths and science should not tolerate lies or fraud in any guise.

As someone who, like Sir Roy, has been involved in medical research for more than 50 years, it is remarkable to me that the occasional scientist or physician is willing to risk his or her career by fabricating the results of experiments. We are all keen to make a breakthrough in medical science as it may ultimately lead to improved diagnosis or treatment of patients suffering from serious diseases. We are also not immune to the respect we may receive from our peers, or very occasionally from the public, if our research is considered a significant advance on what has gone on before. However, fraudulent reports will always be identified eventually.

Sir Roy draws our attention to several fascinating examples of the innate deception that can be seen in the plant and animal worlds. In these examples, deception is an instinctive means of survival or to gain an advantage and it is not a subject for moral and ethical discourse. Humans, in contrast, endowed with a conscience together with the handicap of "Original Sin", have the choice to follow their inborn temptation to cheat or resist the evil option and choose honesty.

Fraudulent behavior is, of course, not confined to scientists. Supposedly respectable bankers and 'financial advisors' or investors have given us multiple examples of this in the recent past.

I have observed several well-known scientists who were guilty of putting 'spin' on their data, over-eager to prove that their hypothesis was correct. Although this is a scientific transgression, it is perhaps understandable, and there is usually at least an element of truth in what is being reported. Sir Roy is concerned with a different category of deception — definitive fraud, knowing that the results of the study are purposely and completely fabricated. With time, if the topic is important enough to warrant others reproducing the experiments, the culprit will be exposed.

One should bear in mind that, on occasions, a reputable and honest scientific team cannot reproduce its *own* findings. Some years ago, I worked closely with an innovative biochemist who developed a storage (preservation) system (in which a special fluid was pumped through the coronary arteries of a heart) whereby we could maintain a pig or nonhuman primate heart alive outside of the body for up to 48 hours, and then transplant it successfully into a recipient animal and obtain long-term graft survival.

This was a remarkable achievement for the era and had required several years of painstaking study by my colleague who had tested multiple potential constituents for his storage solution. So impressive were the results that, as a result of our publications, the biochemist was offered a job in a respected laboratory in the USA. When he tried to reproduce the experimental results there, despite all of his efforts, he could not do so. Yet I can vouch for the success of his system, which in fact we used successfully to store human hearts while being transported from one city to another for transplantation into patients at our medical center. None of us could ever determine why the results could not be duplicated, and yet there was absolutely no fraud or deception involved.

To fabricate research in science requires a positive decision to do so. The decisions we make in life to a large extent govern the roads our lives take. We all make incorrect decisions from time to time, e.g. with regard to paths our careers will follow, the partners we choose, how we invest our money, and so on — some of us more than others. Sir Roy gives examples of some monumentally bad decisions made by military leaders in history — for example, both Napoleon and Hitler in their decisions to invade the vast, inhospitable country of Russia.

Why does a scientist decide to cheat? What influences a hitherto reputable scientist to take a risk that will have so many ramifications for themself and for those with whom they are collaborating? Does he truly believe he will never be found out? This book explores the factors and psychology that might lead a scientist to make such a choice. If the scientist has already had some genuine success, possibly the need for continuing recognition and 'fame' within the scientific community proves overwhelming. Is it a form of addiction — addiction to being praised and academically recognized — that the scientist cannot control?

The scientist's ego and his need for continuing success and attention may bring down not only himself but also his mentor. Sir Roy suggests that the mentor may be a highly regarded scientist, but one who is so overwhelmed by administrative responsibilities and other commitments that he no longer has sufficient time to supervise the work of his juniors as he once did. The deception goes unnoticed until it is too late, and the mentor's reputation suffers from his association with his dishonest protégé. Such a scandal has at times probably prevented a mentor from winning a Nobel Prize.

This book is more than a report of cases of deception in science, interesting as these are. Sir Roy asks pertinent questions, such as what can the scientific community do to identify scientific fraud before it has been published and made available to all unsuspecting scientists, who may subsequently be stimulated to follow a completely fruitless path of research. Given the frailties of human nature, it is unlikely we shall ever live in a world in which scientific fraud is never attempted, but it is possible we can develop methods to identify it at an early stage (before its publication) and thus prevent the dissemination of the fraudulent results. However, how can we prevent fraudulent reports being published in the leading scientific journals? Particularly, when the results appear so convincing to peer-review that they have been accepted as being true — though at times the review of the data may have been less than thorough.

Sir Roy also discusses topics such as how we can help young people to manage expectations for a career in science.

Brief and succinctly written, the book is thought provoking, wide-ranging, and highly readable. It will prove of interest to scientists, physicians, and laypersons alike. I hope you, the reader, find it as stimulating and entertaining as I did.

Contents

Chapter 1
Introduction

1.1 Science Works

Human nature is the driving force which has underpinned the survival of *Homo sapiens* through millennia. Natural instincts enable us to avoid predators, to cope with environmental extremes and to access nutrition. Above all, human nature drives the reproduction of our species which ensures its continuation into the future.

Organising society requires hierarchical control that in all successful nations has deceit intricately and extensively woven into its fabric. These societal structures often rely on the propagation of unsubstantiated myths enforced by those in authority.

Modern science, initiated by curiosity, which is another feature of human nature, has accelerated in its advances in seeking the truth. It does so by reason of the independent reproducibility of each step forwards and the demonstration that a hypothesis is correct or will otherwise be rejected. There is no room for dishonesty in the progress of science.

Unfortunately, like the rest of humanity, scientists are not always able to avoid their natural instincts to submit to the temptation of falsehood. Even if this is ultimately self-defeating. In this book, I suggest that in the search for truth, scientists must maintain higher standards of honesty and rigour than what might be acceptable in other pursuits.

Deception and deceit have evolved in the plant and animal kingdoms in a variety of species. Perhaps the best-known is the cuckoo. This bird instinctively lays its eggs in a nest made by other species. When the chicks from the cuckoo eggs hatch, the chicks then throw out the eggs from the rightful builder of the nest. This is fraudulent behaviour, which is not learned but is instinctive upon birth.

Many bird species, especially members of the Corvid family, exhibit remarkable intelligence despite their small brains. The tool making crows of New Caledonia can fashion precision instruments to extract insects from their nests. The Drongo bird of the Kalahari Desert can mimic the alarm calls of meerkats and other creatures to frighten these competitors away from their food, so that the drongos can then steal it for themselves.

A recent *Nature* publication described fourteen examples of rove beetles which have evolved to mimic army ants. This enables the beetles to enter the nest of the army ants with impunity before eating the younger developing ants contained within. Deception is therefore an instinctive element of survival amongst many creatures within the animal kingdom.

Possibly, in humans, the temptation to cheat may also have evolved with an instinctive basis as part of what has been described in some religions as Original Sin. The original sin was derived from the bad root, Radix Mali, which according to Judeo-Christian doctrine, was conferred by Satan onto Adam and Eve.

The concept that humans have a burden of sin at birth, was championed by Saint Augustine of Hippo, 354–430 AD. He was known to have been a colourful character in his youth and after entering the church wished to postpone his own deliverance from evil in his prayer, "*Please God make me chaste, but not yet*".

Augustine stated that language is not given to man for mutual deceit, but for the exchange of ideas. According to Augustine, anyone who uses language to deceive misuses it, which is therefore a sin. The opposite sentiment, however, was espoused by the French diplomat Charles Maurice de Talleyrand. In conversation with the Spanish ambassador Eugenio Izquierdo in 1807, Talleyrand is alleged to have said, "Speech was given to man to hide his thoughts."

Homo sapiens would appear to have the option of whether or not to indulge in deceit. Certainly, throughout history there are numerous examples of deceit, which vary from mild peccadillos to deceptive behaviour that changes the course of history.

Cheating appears in literature and poetry from the first evidence available. We are continually subjected to lies and deceit, a total way of life in dictatorships but one that seems to be becoming a *modus operandi* in countries that would claim to be democratic. The statements of politicians are frequently so far from the truth that the man in the street often ignores them as normal misinformation.

In the scientific context, an error may not initially be deliberate, but once an experiment has been misinterpreted, the rot can set in and lead originally honest scientists to deceive.

In this book, I will examine three examples of serious fraud in medical and biological science. I will try to seek the origins of each decision to falsify, bearing in mind that often the greater the fraud the more certain it will be rapidly uncovered and the perpetrator exposed as a cheat.

The three cases involve cultured cells. They had disastrous consequences and led to deaths and disgrace of internationally famed institutions. These include: the Memorial Sloan Kettering Cancer Institute in New York, the Harvard Medical School in Boston, the Riken Institute in Kobe and the Karolinska Institute in Stockholm, which is responsible for awarding the Nobel Prizes in Medicine.

These three cases of scientific fraud all feature aspects of deceit which are prevalent in human society more broadly. We are frequently subjected to misleading or "fake news" in the media and in general, society disapproves of cheating, with some forms considered to be criminal practice.

In human endeavours, sometimes the cheat prospers and avoids detection. In politics, increasingly brutal methods are used by despotic rulers to ensure they can continue to cheat with their profitable corrupt practices. Unfortunately, lies and fake news often escape the moral filters that one would expect to be present in a democratic society.

In our attitudes and behaviour, most humans probably abide by the Dickensian code to 'live and let live', (an old proverb which appears in Dickens' 'Bleak House'). An important minority, however, would go out of their way to pursue a life devoted to helping others who they perceive as being underprivileged or persecuted.

On the other hand, there are a few misanthropes, who seem to have no conscience holding them back from criminal and violent actions against others. I have drawn a bell curve to roughly illustrate this categorisation of human nature (Fig. 1.1). A shift to the right of the curve can occur with terrible results when a whole nation, with a cultural history of hundreds of years, is incited to hatred by a charismatic popular leader.

Science should be different, but deceitful motives have a variety of origins, many being difficult to understand. Shame and loss of career prospects should be a deterrent

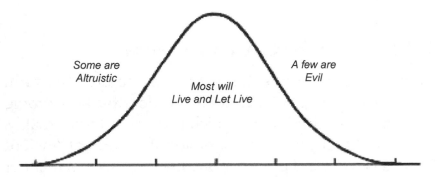

Fig. 1.1. Schematic bell curve of human nature variation (RYC).

for most scientists who are tempted into corruption, but on the other hand, serious penalties for most crimes are seldom used to punish scientists who cheat.

Examples of deliberate falsification in science may be motivated by a wish to unsettle and undermine the credibility of orthodox scientists in the field. This is usually a well thought-out fraud. For example, the claim of finding the *'missing link'* in human evolution was a plot hatched in 1908 by an amateur archaeologist, Charles Dawson, who claimed to have discovered teeth and fragments of jawbone and a distorted human skull, which were at least 500,000 years old, in a dig at the East Sussex village of Piltdown. It wasn't until 45 years later, with the help of new techniques of bone dating, that the bones were proven to be a mix of orangutan and modern human bones. The 'Piltdown Man' scandal misled scientists for decades, an unusually long period of time for such a major fraud to last.

The state of mind of scientific cheats who deliberately make falsifications, which they know could be discovered, could be regarded as a form of hubris, which seems to affect human nature from the least to the most intelligent.

1.2 Hubris in History and Science

Hubris is a common feature of human nature. The term originates from Ancient Greece and denotes an extreme pride and self-confidence which is divorced from reality. It was said that arrogance annoyed the gods of Ancient Greece who would punish individuals acting in a manner which was too vainglorious.

Hubris has been a present aspect in many of the most famous episodes of human history. Napoleon and Hitler attacked Russia without due consideration of the Russian winter. Great Britain, Russia and NATO attempted to defeat the Afghans without contemplating the history of two ignominious military failures that Great Britain suffered against the Afghans in the nineteenth century. A similar fate in the 20th century befell the Russian invasion. Japan decided to attack Pearl Harbour in 1941, completely blind to the potential strength and outrage that would mobilise America. Like these historical examples, scientists too are not immune to hubris.

Short-term advantages can be gained by scientists who persuade prestigious journals to publish their work. This can lead to individual promotion, the award of grants in a poorly funded and competitive field and sometimes personal financial reward. Of course, ambitious scientists are vulnerable to the temptations familiar through human history of greed, jealousy, revenge and theft. Therefore, it is not surprising that there have been many examples of these forms of scientific corruption.

I have chosen to examine three case studies in this book. The first is the skin graft claims of William Summerlin. The second is the acid culture production of cells resembling embryonic stem cells in Harvard and Kobe, by Haruko Obokata and the third is the disastrous plastic tracheal transplants in Stockholm and Russia by the surgeon Paolo Macchiarini.

In the course of trying to understand these three examples of scientific fraud, I have been fascinated by the way in which the initial deception has been defended not only by the perpetrator but, extraordinarily, by senior scientists and university administrators. Through their support of the deception, they become seriously implicated in delaying the unmasking of the cheat.

I am interested in examples of young, initially honest scientists who may ignore the fact that they could not repeat their experiments. Often, they were not prepared to examine the possibility that the hypotheses and design structure of their experiments may have been flawed from the beginning.

There is a danger for scientific investigators who imagine that his or her experiments will lead to fame, prizes and rapid promotion. In the light of apparently positive early results, expectations can be raised and alternative explanations dismissed completely in their minds, which have become biased.

Having committed their erroneous observations and conclusions to publication, they start to dig themselves into holes, which become deeper as more attention is paid to their results. This sad and dismal process is quite common and not difficult to understand. It is almost always the result of inappropriate personal supervision and often occurs in large and famous institutes, whose heads are engulfed in university and political administration that takes up much of their available time. This alone is probably an argument against oversized scientific departments.

There is an extensive literature on cheating and misconduct in science and medicine published as academic texts. There is no doubt that fake science has always existed and it unfortunately undermines the vocational ethics of honest scientists. Probing for the truth is difficult; there is a general wish to be charitable in attributing the benefit of the doubt when distinguishing genuine mistakes from the sheer effrontery of cunning and barefaced lying.

In 1997, Michael Farthing editor of the journal *Gut* led a new attack on scientific misconduct with the formation of a UK Committee on Publication Ethics, COPE. This distinguished group of journal editors, lawyers and Professor of Ethics, Ian Kennedy, defined important issues in the detection and elimination of fraud in medical and scientific journals. Nevertheless, cheating, often on a serious and worrying scale still persists. Most medical scientists have trust in their colleagues and are naturally disinclined to challenge suspected false claims until the evidence is overwhelming.

In this small book aimed at readers with a scientific background, students contemplating science as a career and curious non-scientists, I have focussed discussion on the restricted field of medical bioscience with which I am familiar. Connected to the cases examined, I have tried to explain how there is a tendency in most humans to be 'economical with the truth' and lie.

The acceptance of downright lies by experts who should know better is not uncommon. They often demonstrate a forceful defence of the fraud, revealing the gullibility of some of the most respected specialists.

I have illustrated a similar weakness in a consideration of examples in fakery and false assessment of originality in art and the world of wine. The sacrosanct status often conferred on experts in these subjects can be astonishing.

I argue that science has no place for such dishonesty, but as in all walks of life, scientists can be led into temptation to cheat, which can be difficult to resist. The consequences may seem to be trivial, but in the mega-scandals considered in this book, not only does fake science hold back advances, it can also shake faith in the integrity of scientific enquiry. Science works because it depends on independent reproducibility of experimental results, or rejection if this cannot be demonstrated. When fake science comes to light, it diminishes the faith of the general public in the scientific process and the majority of honest scientists within it.

Applied science frequently improves the lives not only of humans but, if used with thoughtful compassion, should benefit all living creatures. Scandals of fraud in science cause shame, harm and death. To reduce the incidence of fake science would be immensely difficult. I have made a few tentative suggestions towards this objective, which I would not claim to be original, but are worth revisiting in the light of recent extremely harmful mega-frauds considered in this book.

Chapter 2
Faking in Plants and Animals

"Will you walk into my parlour?" said the spider to the fly "Tis the prettiest little parlour that you ever did spy. The way into my parlour is up a winding stair, and I've many curious things to show when you are there."

Mary Howitt, 1829

2.1 Deceiving Plants

The instinct to deceive is a successful evolutionary stratagem in some species of plants and animals. For these species, deception carries no moral or ethical baggage but is just another evolutionary development in the 'survival of the fittest' which helps to ensure their continued existence.

The Bee Orchid, for example, has developed a "Honey Trap" which attracts male bees to a decoy petal that resembles a desirable female bee. The flower deposits pollen on the unsuspecting male bee, priming it to further propagate the orchid species when it is deceived by a second seductive flower (Fig. 2.1).

The Venus plant, a native of the Americas is armed with leaves adapted to function as efficient flytraps. These

© Tim Wilkins

© Christian Ziegler/Minden Pictures

Fig. 2.1. Collage of Bee orchid with a petal resembling a female bee and Bee orchid with a bee seduced by the petal — Philip Ball (IPD).

specialised leaves have vicious looking teeth-like spikes, which clamp down to trap the unsuspecting insect that has been attracted to the nectar odour emitted by the leaf (Fig. 2.2).

Pitcher plants flourish in tropical humid areas in Asia and the Americas. They resemble a jug for holding water and are formed of modified leaves. Some have configurations that are beautiful (Fig. 2.3). Like the Venus plant, they emit a sweet nectar-like odour, which attracts insects.

Fig. 2.2. Venus plant with captured insects collage — Philip Ball (IPD).

Fig. 2.3. Pitcher plant with insect falling into the trap (IPD).

During falling rain, the rim of the trap becomes very slippery and insects alighting on it will lose their footing and fall into the lethal digestive juices within the cavity. In some species, a rapid, highly dynamic trap door mechanism will snap shut eliminating any possible escape route.

These highly sophisticated evolutionary developments underline that deception is simply a natural strategy for survival in the plant kingdom. When it comes to cheating to gain an advantage, plants, unlike humans, are not constrained by the moral concern of a 'conscience'.

2.2 Deceiving Animals

Many different forms of deception occur in the animal kingdom too. A galaxy of strategies have been used by fish, birds, non-human primates and insects so that weak but intelligent males can succeed in mating with females. Subtly, these males use the advantage of their small size to outflank larger alpha animals, with speed of action to get in first. Some, for example, have developed sexual mimicry of the opposite sex. The elaborate behaviour of the cuckoo fools other bird species into caring for the cuckoo chicks instead of the chicks of the genuine nest builders and egg producers (Fig. 2.4).

The Anglerfish (Fig. 2.5) has an illuminated rod on its head that lures prey to within reach of its huge jaws which are equipped with vicious razor-like teeth. The praying mantis and many species of spider eat their mates after sex to secure extra nutrition for their newly fertilised progeny (Fig. 2.6).

Fig. 2.4. Reed warbler feeding a cuckoo in its nest that it has mistaken for a chick of its own (IPD).

Fig. 2.5. Angler fish with enticing beacon on the top of its head to attract prey (IPD).

© Gizmodo: George Dvorsky

Fig. 2.6. Female Praying Mantis after mating has just enjoyed a meal of the head of her mate.

Some of these examples can be regarded as mimicry rather than cheating and the naturalist Henry Bates described them as simply harmless developmental traits to avoid potential hazards (Bates, 1862, Linnean Society).

These are just a few of many examples of cheating which have developed over millions of years in the natural world. These evolutionary advantages which are gained through deceptive behaviour, are also present in human society and amongst the scientists who work within it.

Chapter 3
Cheating in Human Society

3.1 Deceptive Instincts in Human Nature

In the evolutionary struggle for human survival, the fittest may gain advantage by lying. Babies from six months of age learn that they can benefit by crying and laughing to gain attention. This experience may be adopted as a useful strategy to be employed deceitfully without any causal pretext. The instinct of *Homo sapiens* to dissemble and cheat may be similar to that of plants and animals, but with the important additions of the moral and ethical constraints of conscience.

Deception and lies figure prominently in all co-operative groups of humans extending from hunter-gatherer tribes to major national communities that consider themselves to be highly cultured in their civilisations. Most societies seem to need some form of supernatural belief or religion to organise society in a controlled manner usually regarded as 'civilised'. However, these beliefs can be deceptive and the perception within societies that they themselves are 'civilised' is often revealed to be a terrifyingly thin veneer.

When natural disasters like earthquakes and hurricanes strike, those afflicted can be deprived of the normal protections of shelter. In these scenarios, individuals often suffer desperate hunger, thirst and infective diseases and these circumstances can lead to an increasing temptation to loot, rob and even murder. That being said, disaster is not a necessary prerequisite for anarchic violent behaviour. In 1969 in Montreal, the mere temporary removal of the restraining influence of the police force, which went on strike, was rapidly followed by mayhem and murder.

1969: Montreal's "Night of Terror"

The story

> Montreal is in a state of shock. A police officer is dead and 108 people have been arrested following 16 hours of chaos during which police and firefighters refused to work. At first, the strike's impact was limited to an increase in the number of bank robberies. However, as night fell, a taxi drivers' union seized upon the police absence to violently protest a competitor's exclusive right to airport pickups. The result, according to a CBC Television special, was a "night of terror". Shattered shop windows and a trail of broken glass evidenced the looting that had erupted in the downtown core. With no one to stop them, students and separatists joined the rampage. Shop owners, some of them armed, struggled to fend off looters. Restaurants and hotels were also targeted. A corporal with the Quebec provincial police was shot and killed at the garage of the Murray Hill limousine company as taxi drivers tried to burn it down (CBC Archives).

Montreal soon recovered when the police presence was restored. However, the events which occurred that night highlight that when the restraints of civilisation are removed, many humans cannot resist their survival-of-the-fittest instincts to deceive which often lie just beneath the surface of 'civilised' society.

3.2 Perception vs Belief

I wouldn't have seen it if I hadn't believed it.
Marshal McLuhan (Canadian Philosopher)

The Emperor has no Clothes
Hans Christian Anderson (Danish Author 1837)

Ignoring the Elephant in the Room.
Ivan Krylov (Russian Writer 1814)

The two classic stories of the naked emperor and the elephant filling the room refer to the natural but dishonest tendency to deliberately ignore important but potentially dangerous or unwelcome truths. In both stories, even though a truth is perceived by the majority of those present, the expression of what these individuals actually believe is prevented by social restraints. Beliefs can often be held with intense determination without adequate factual support. Perceptions meanwhile are typically formed by things which we come in contact with using the five senses of sight, hearing, smell, taste and touch. However, the differences between perception and belief can complicate how we as humans perceive 'truths'.

For example, sight is powerful, however, it is not necessarily an objective truth. The perception and interpretation of images focused on the eyes' retinas are distributed and analysed in different parts of the brain. Putting together and memorising images can vary in accuracy and completeness and therefore the final perception of an image will be greatly influenced by the context and personality of the individual who perceives it.

If I enter an operating theatre, my vision is focused on the progress and successful outcome of the operation. A senior operating nursing sister would be concerned with the array of instruments and make sure they are the correct tools for the job, arranged in the right order. An anaesthetist would wish to determine that there was good lung ventilation, heart function and the safe and effective level of anaesthesia. Each is viewing the same scene. A journalist reporting on our transplant surgical programme for the media, walked into the room where I was operating, scanned the scene with horror and immediately fainted!

What we think we perceive can also itself be deceptive. A master magician and conjurer can repeatedly control the perception of his audience by speed of his sleight of hand to distort or eliminate conventional image perception. Therefore, the quote at the beginning of this sub-chapter could be rephrased to *"because I saw it with my own eyes I believed it"*. It is not surprising that the recall of events, especially when presented as evidence in court, is often shown to be unreliable or just wrong. Perception and belief influence all aspects of human society and importantly they complicate what different people perceive as the truth.

3.3 White Lies

Closely linked to the difference between what we see and what we believe, is the tendency within human society to tell 'white lies'. Social interactions can be difficult or impossible without the use of white lies and the mincing of words. However, when we tell white lies we are expressing a sentiment which is normally different to both what we perceive and what we believe. Consider a hypothetical dialogue between partners:

> She.... "You look so young and handsome!"
> He...... "You look lovely in your new dress!"

Both smile at each other and hold hands. If, however, the veneer of the kind white lies is stripped away, the dialogue could be rather different:

> She...."You are going bald with bags under your eyes and there are food stains on your shirt."
> He....."That teenage dress is far too short and with your wrinkly appearance you resemble mutton dressed as lamb."

White lies are another example of how civilized human society actually depends on falsity for some of even the most basic of social interactions. That being said, the truth can cause hurt and resentment, so much the better the more gentle 'white lies'.

3.4 Lying in Politics

One area of human society where lies are considered to be particularly prominent is political discourse. Politicians and bureaucrats will often deceive the public by talking with "forked tongues" and using words which mean the opposite to their true intentions. For example, they might use the terms "clarity" and "transparency" to present acceptable policies, when in fact they plan to act in a totally opaque manner to introduce unpopular decisions. Similarly, the phrase "to be honest" is often followed by a bare-faced lie.

Modern science provides new and powerful tools through which political pressures can be applied. The rise of social media now means that political messages can spread at a frightening pace, irrespective of the veracity of their claims. Social media disseminated through the internet was partly responsible for the rapid rise of the Arab Spring.

The Arab Spring started as an angry response to a totalitarian regime in Tunisia. It began when a poor street vendor was deprived of his living and was evicted from his modest stall. Unable any longer to support his family he set himself on fire in protest to the hardships he had experienced. The popular outrage and a demand for a change to a democratic government reverberated through-out the world. It was initially perceived as a progressive movement, however sadly, the Spring rapidly changed to an oppressive winter. In countries that had seemed to be experiencing joy and hope, these feelings were wrenched from their grasp, with the imposition of a violent repression, even more brutal than before the Spring.

The Arab Spring is just one example of how hope for political change can deceive large swathes of a human society. Importantly, it also highlights how modern social media now accelerates the dissemination of political rhetoric, true or false, amongst the general public.

The deliberate manipulation of the democratic spirit and process can be a cunning strategy to gain power. In some cases it can be used to establish a dictatorship that will eliminate all types of opposition.

The aspiring despot brands all legitimate and reasoned opposition politicians as traitors and terrorists to be punished or banished. If they can be identified ethnically and stand out as separate religious or tribal minorities they may be labelled as enemies of the people and subjected to vitriolic hate rhetoric, personal violence and may even be eliminated through ethnic cleansing. Thus, from small-scale political lies to totalitarian repression deceit is an ever present characteristic in the politics of human societies. Even in countries which label themselves as democratic, deception and cheating are seldom far away from the political process.

3.5 Forgery in Art

Cheating even permeates humankind's leisure pursuits. Every sport appears to be rife with deceptive practices from the use of illegal performance enhancing drugs to disguised fouls and theatrical 'diving' at non-existent tackles. Similarly, forgery in the world of art remains a pervasive problem. The fakes are often so skillful that the experts and dealers are completely taken in, much to their subsequent fury. However, it can be difficult to know where to draw the line as to what is flagrant cheating as opposed to acceptable derivation.

Almost all the original paintings and most of the sculpture of Ancient Greece comes to us via Roman copies, some probably executed by Greek artists living in Rome. Whilst the skill of these copiers was venerated by the ancients, in modern times, art copiers are set upon by those who have been duped. These individuals, who typically have enormous financial resources available to them, will hire the most skillful and expensive lawyers to secure conviction and retribution.

An extremely talented artistic forger, Eric Hebborn described in his beautifully illustrated book *Drawn to Trouble,* how as an impoverished art student he managed to get a part time job with a self-styled art restorer. Hebborn was presented with a totally blank 18th century canvas and told to restore it to its pristine splendour as a painting of Dutch warships by the younger Willem Van de Velde. He was given pocket money to visit the Royal Maritime Museum to hone his expertise on the technical skills of Van de Velde. Eric Hebborn's 'restoration' must have been excellent, since it is thought to have been proudly exhibited for many years in a national art collection until Hebborn was unmasked as a forger.

Hebborn also expressed in his book the joy he experienced when visiting an esteemed Old Bond Street Gallery to see some delightful ancient ink drawings. He felt they looked far more beautiful on the walls of the gallery than when he had last seen them in Rome, drying on his washing line.

When I came across an image of a beautiful young Florentine girl, "La Bella Principessa", in a newspaper (Fig. 3.1), I assumed that it was a work of Leonardo da Vinci that I had never seen before. I then read that it was claimed to be a forgery by the self-taught artist Shaun Greenhalgh. He had done the painting in a shed in his modest back garden. His model was "Bossy Sally", a checkout girl at the local supermarket. He described in detail how he produced the painting, including the procurement of the velum, the pigments he used and how he gave the impression of a left-handed artist. He sold the painting to a friend for £80.

© Shaun Greenhalgh

Fig. 3.1. La Bella Principessa — Original Leonardo da Vinci or forgery by Shaun Greenhalgh (IPD).

The painting then disappeared mysteriously in the hands of dealers and collectors and eventually emerged at auction valued at $150 million. Shaun Greenhalgh pointed out that the version that had reappeared had been modified by others, but was still his original. Attribution of the artist of this painting has been subject to considerable scholarly dispute to which I am not qualified to make any contribution.

Poor Shaun Greenhalgh was sent to prison for four and a half years, where he wrote a fascinating book about his painting and sculpture forgeries, *A Forger's Tale*. He was treated harshly and, many would feel, completely out of proportion compared with the sentences used to punish those guilty of violent crimes.

The cases of Greenhalgh and Hebborn highlight how the line between what human society deems as acceptable derivation and what is outright plagiarism is not always clear. This line has certainly changed throughout human history. In Roman times Greenhalgh and Hebborn's ability to mimic the works of great artists would have been lauded. If they had been born in 15th century Florence, their extraordinary skill and work ethic probably would probably have won them an apprenticeship to one of the famous Florentine artists. Their paintings and sculptures would eventually have been recognised as those of great painters in their own right, or at the least their works would have been attributed to their masters or masters' schools. Greenhalgh would certainly not have languished in prison with the time and the inclination to write his extremely entertaining book.

From white lies to politics to art, fraudulent behaviours permeate human society as they do the natural world. Scientists are not immune, and since the birth of modern verifiable science, fraud has been revealed in all its different branches, including the most distinguished scientific institutions. Unfortunately, this undermines the credibility of science and the respect for scientists amongst the general public.

Chapter 4

The Unverifiable Beliefs which Underpin the Organisation of Human Societies

"Those princes who do great things, have considered their word of little account, and have known how to beguile men's minds by shrewdness and cunning."

Niccolò Machiavelli

We have seen clear and convincing examples of elaborate and successful deceit in both the plant and animal kingdoms. The extensive modifications of structural anatomy and physiology have evolved over millions of years in some plants and animals to gain essential survival advantages. Similar changes are not as obvious in the anatomy and physiology of *Homo sapiens*, however, our behaviour would suggest that the human brain also contains an instinctive evolved function that frequently involves fraudulent activity.

Once humans had advanced to the point where they could communicate with each other by signs, images and speech, they tended to form family and tribal groups. Their ability to flourish depended on some organisation with leadership and a hierarchy involving the division of duties, such as hunting and gathering, fighting, farming, childcare and cooking.

In the evolutionary hinterland between non-human primates and ourselves, the primatologist Frans de Waal, and others, have observed co-operation, empathy and even grief. These features are especially well developed in the gentle Bonobo chimpanzee, however, their lack of language may have held back the development of what we might describe as a moral awareness and conscience.

The Piraha, an Amazonian tribe has been studied by Daniel Everett, who arrived as a missionary. After living with the tribe, Everett converted to the more simple way of life of the Piraha. They hunt and gather without having words, numbers or a religion. They live by the day, with no obvious consideration of the past or future and seem to be carefree and happy (Daniel Everett, *Don't Sleep, there are Snakes*, Vintage Books, 2009).

Most conscious self-aware humans, with few exceptions, such as the Piraha, yearn for explanations of mystical matters that are the foundations of all religions. Curious individuals in separate scattered communities started to wonder if there were answers to the basic questions of *"Why are we here?"*, *"Where did we come from?"*, and *"Where are we going?"*

Confronted with these worries, humans have always sought explanations to allay their concerns and provide themselves with hope and reassurance. There was a natural tendency to turn to the group leader, who of course could not provide answers.

Nevertheless, to maintain prestige and respect, these leaders sometimes created myths which were often very elaborate and fanciful in content.

The structures of all societies rest on unverifiable myths and superstitions, which were usually comforting, reassuring and acceptable. They needed to be tolerated; otherwise, the integrity of the community would break down.

Creeds became consolidated separately in different tribal groups often with complicated ceremonies involving dancing, singing and drumbeats, especially connected to sex, marriage and death. Usually separate doctrines proclaimed their own deities to be worshipped. The sun, moon and stars were well-defined icons that could be understood by all. Divinity also was bestowed on powerful despots themselves for example the early Pharaohs and some Roman Emperors.

Other communities chose to make gods of revered animals and natural phenomena, such as mountains, rivers and oceans and even specially constructed sculptured idols. In the ancient cave paintings of our forebears, which date back up to 35,000 years in southern Europe, these early *Homo sapiens* depicted animals that they hunted or hunted them. The images are thought to have had important mystical and superstitious properties, conferring safety and successful hunting for the community.

The concept that man is fashioned in the form of an unseen God has been very successful and continues after thousands of years. This developed into a tendency amongst tribes and nations to fiercely defend their own particular religions. If individual creeds were challenged by those of another group, it often became a *casus belli*, even if the dispute might appear to an outsider to be only a minor difference of interpretation. Revenge by torture, rape, murder and enslavement of subjugated foes were justified in the name of fulfilling the doctrines of religious creeds and sadly, this continues. This pattern of behaviour was easily adapted to the political management of societies, with the establishment of monarchies by strong men and women determined to keep power in the family bloodline.

A variety of systems, often successful for many years, such as Soviet Communism, demanded blind obedience with no dissent permitted. This is still the pattern of authoritarian dictatorial regimes. Orders are given irrespective of whether or not they are based on truthful information. The obedient population becomes used to a succession of lies, which are embraced and accepted in order to survive. Even in so-called democratic societies, opposing politicians increasingly canvas for votes by dishonestly promising favours that they know they cannot provide.

This culture of dishonesty permeates almost all aspects of society.

4.1 Truth and Political Correctness

"In a time of universal deceit, — telling the truth is a revolutionary act."

George Orwell

Over time, religious and political creeds have been pruned of the branches that are no longer considered valuable to them. Whilst at the same time, new shoots are added to fit the ever-changing demands of modern society. So-called politically correct speech has been extensively adopted and accepted by Western society. However, this growth is not always a positive development for the progress of science through free academic discussion.

Recently, political correctness has eliminated discussion and debate in prestigious universities. Similarly, academic visitors have been denied access to speak, being subject to no-platforming. Many thoughtful scholars are scared to object for fear of persecution on social media, and losing their jobs. This is a very worrying development that appears to be getting progressively worse. In Orwellian thinking, PC could be interpreted as *politically incorrect* and social media might be better understood as *anti-social media*.

The culture of science should be quite different, as evidence is sought to support or refute a particular hypothesis. However, unfortunately, the political environment can still influence acceptance of the evidence, even when the data may be clear-cut. Galileo for example had to face the Inquisition of the Church for very gently suggesting that the earth travelled around the sun. This contravened the Aristotelian dogma which had previously been accepted and Galileo was lucky to avoid execution.

Escaping a similar fate, Copernicus could perhaps be considered fortunate that his book on the earth orbiting the sun was published in 1543 on the day he died. Friar Giordano Bruno's ending was tragic; he was burnt at the stake in Rome for the same heresy.

Certain obvious highly relevant matters, such as the apparently unstoppable increase in the human population, are scrupulously avoided in most important international policy forming meetings, for fear of offending some religious and political organizations. Political correctness and the evasion of uncomfortable truths can therefore undermine the truth in science and in human society as a whole.

4.2 Science and the Nature of Truth

"Science is incomparably the most successful human activity human beings have ever engaged upon."

Peter Medawar (Nobel Prize Winner)

Whilst I do not wish to get entangled in the perennial philosophical question of the nature of truth, the concept of this book implies that truth in general can be differentiated from truth in science. The word "truth" can be applied to certain manifest and axiomatic beliefs, for example, all humans are mortal, boiling water on the skin will burn, newborn babies require care and nourishment and each day the sun will rise and set.

Science can help prove 'truths' such as these. For example, scientific observations and measurements can predict that sunlight will be temporarily excluded at a certain time and place when the path of the moon passes between the sun and the earth. This we know commonly as an eclipse. The strength of the scientific method will be enhanced if predictions such as these are fulfilled, however, if they are not then the scientists must go back to the drawing board. Science can therefore provide certainties where the unverifiable beliefs which underpin the organisation of human societies cannot.

Chapter 5

Why Science must be a Verifiable Truth

"The only purpose of science is to find the truth."

Allan Kirk (Professor of Surgery Duke, University)

In the history of learning, science is the new kid on the block, and quite different to other traditional subjects. Unlike other subjects, the scientific method demonstrated that verifiable results from experiments actually worked and did not rely on mythical concepts or unverifiable opinions.

The dawn of science began with the major contributions of ancient Greek mathematicians, Archimedes, Pythagoras and Euclid. They used logic, written concepts and above all measurement, which were all essential for modern science to flourish. Thomas Korner, professor of mathematics at Cambridge, noted that "Euclid sits at my elbow whenever I construct long mathematical argument". Korner also rightly pointed out that the torch of Greek mathematics passed through Constantinople and Baghdad and onwards to a Europe beginning its scientific Renaissance. Inventions of new instruments such as the microscope, telescope and especially the printing press helped to bring science into a new era of progress and towards its modern form.

Despite the overwhelming number of examples of lies and deception all around us, even for those living in superficially tolerant and liberal democratic societies, I would advocate that the nature of science is of a different order to most non-scientific interactions. Scientific advances require verification by being reproducible by independent workers and even then, the scientific community only accepts claims of major advances if application of the new data actually works in practice.

Thus, will a new antibiotic really eliminate a virulent bacterium without toxicity and cure a patient suffering from a fatal disease? Does a new alloy provide added strength and durability when used in the construction of an aircraft's wing? Affirmative observations will in the short term have only varying degrees of tentative acceptance. The rigorous discipline of science must therefore ensure that false claims are detected.

Fake science is inadmissible and will always be found out eventually. Usually the more important the subject and the bigger the fraud the quicker the deceit will be unmasked. A great deal of harm and sometimes tragedy may in the meantime result from the cheating.

5.1 The Use of Science in Unmasking Fraud in Art and Wine

Because the scientific method demands a superior order of honesty to that which is to be found in most other pursuits, it is used as a measure of truth in other fields.

For example, science is used in various capacities to prove and disprove the claims of authenticity in the worlds of art and wine.

In her book *The Scientist and the Forger* (2015, Imperial College Press), Jehane Ragai describes the rapid advances that have been made in physics and chemistry in developing new and increasingly accurate instruments to detect fraud in art. For example, a dispersive Raman spectrometer can be used to analyse the exact types of pigments used in the paint of a particular artwork. This analysis can help determine if these were the pigments commonly used and available to certain artists during the time that they were said to be painting.

Although some of the machines are extremely expensive, the financial consequences at stake can be huge. An original artwork can sell for hundreds of millions of pounds, whilst a convincing but unattributed fake may be valued in thousands or less. Thus, machines such as these are extremely useful in the verification process regarding whether an artwork is authentic or not.

On the 15th January 2017, a painting attributed to Leonardo da Vinci, "Salvator Mundi", depicting Christ as Saviour of the World, was sold in New York for the record price of $450 million. It had suffered from extensive restoration and had been sold in the 1950's for £45 as it was considered to be a copy. Now its provenance of attribution is still disputed. The relevance of the money exchanged to the quality of the art is clearly ridiculous. The auctioned painting has already been criticised for the indistinct gaze of the eyes, which are dull, the neck that is too thick and the lack of refraction that should be apparent through the crystal sphere that Jesus is holding. All are said to be impossible errors for Leonardo to have made.

It is not therefore surprising that no expense is spared, if the potential profits and losses can be so great. In addition to the science, attribution also depends on considerable weight being given to the opinions of art experts, whose independence is sometimes open to question. The final decision can be made in a law court by a judge who has no special expert knowledge or even interest in art. When there is a legal conflict in an art law case, the judge may be influenced by the eloquence and skill in the advocacy rather than the artistic knowledge of the lawyers acting on behalf of each side. In these cases, science as an objective truth can help judges come to the most informed decisions.

Science is similarly employed to establish the originality and veracity of certain wine vintages. The discovery that the fermentation of grapes and grain created alcoholic drinks that could prevent the spread of diseases, but also relaxed the mind for convivial human interactions, has brought joy to many.

The appreciation of wine figured heavily in both the Greek and Roman cultures and the fake attribution of the provenance of rare wine has been a goldmine for crooks ever since.

These fraudsters take advantage of the overblown self-assessment of the tasting abilities of gullible people with deep pockets. The label on the bottle, easy to substitute, can completely distort the appreciation of the product. The cork vs screw cap dispute now seems to depend mainly on the sound of the cork popping, rather than anything to do with the taste buds in the mouth (Charles Spence, Oxford). Ancient bottles with matching labels may be revealed as almost pure vinegar, although to acknowledge this in the company of experts might result in a dreadful loss of face.

In the saga of *The Billionaire's Vinegar* published in 2008, Benjamin Wallace, an expert writer on wine, tells the story of a fake bottle of 1787 Château Lafite Bordeaux, allegedly ordered by Thomas Jefferson, sold at auction in 1985 for $156,000. What followed was the sale of more similar bottles and the financially rewarding continuation of the deceit. This snowballed into an elaborate trail of corruption, collusion, lies, threats and lawyers. Eventually after 20 years, modern science was recruited from both sides to analyse the provenance of the labels, bottles and fluid inside the bottles. The results led to some clarification of what had happened and the amazing details of how many gullible people had been sucked into the scandal.

Frauds in art and wine as well as other human pursuits like sport can depend on science to help find the objective truth. It is for this reason that public trust in science is so important and that cheating in science can be so damaging.

5.2 The Fertile Pastures of Modern Science for Fraudsters to Exploit

Whilst science can provide certainty in other human pursuits, the discipline itself is by no means immune from cheating and deception. There has been intense scientific interest in stem cell research over recent years, due to the fact that stem cells have the unique potential to differentiate and renew themselves into specialised cell lines. These cell lines can then be used to make a number of different tissues and organs in the body. Whilst stem cells have been used as a valuable treatment for some malignant and benign blood diseases, they have also proved to be fertile soil for fraudsters to exploit.

Undifferentiated mesenchymal stem cells (MSC's) can be obtained by direct puncture and aspiration from bone marrow as well as from other sources such as umbilical cord tissue. Following culture in the laboratory, some of these cells can be differentiated into fat, cartilage, bone and with advanced techniques into more specialised cells. Both the numbers of cells and their potential for differentiation is unlimited.

With this broad range of possibilities, unscrupulous corrupt practitioners have offered treatment with MSC's to desperate unsuspecting patients. These so-called new therapies are frequently offered without experimental data to support them. They sometimes can have no beneficial effect and can even impoverish or endanger the vulnerable and trusting patient.

Hisashi Moriguchi, a Tokyo University nurse with no scientific credentials, claimed to have cured heart failure with iPS (induced pluripotent stem cell) treatment, working in collaboration with Harvard Medical School and the Massachusetts Institute of Technology. When challenged with the provenance of his claims, all three august institutions denied any knowledge of this audacious deception. Later he admitted the fraud, yet at a press conference, he attempted to justify his deceit. (Nature News and Comments, November 2012).

In South Korea Professor Hwang Woo-Suk, successfully cloned a dog by nuclear transfer with widely published endearing images of the beautiful Afghan Hounds (Fig. 5.1).

© Seoul National University

Fig. 5.1. Cloned Afghan hound by Professor Hwang Woo-Suk. (IPD). © Seoul National University.

He later announced untruthfully that he had also cloned a human embryo. The worldwide uproar that ensued was aggravated when it was learned that he had used human eggs, procured not entirely with goodwill, from members of his staff.

He was forced out of his lab in disgrace and was given a 2-year suspended jail sentence. He reappeared not long after in another lab to continue his experiments. Undoubtedly, he is a skillful and ingenious scientist, but his unethical behaviour leaves important questions to be answered. (Nature News and Comments, January 2014).

The Holy Grail of aspirations for researchers in organ transplantation is to develop a safe and repeatable technique to produce specific immunological tolerance to grafted organs between unrelated people. In other words, to be able to transplant an organ that will function long-term without the need for immunosuppressive therapy. It was therefore not surprising that a claim to have achieved this for kidney grafts by a relatively straightforward protocol in monkeys was greeted with great acclaim. The leader of the experimental team was a popular and well-respected scientist, Judy Thomas, working with Juan Contreras in the University of Alabama at Birmingham.

Unfortunately, other distinguished workers in the field were unable to repeat the Alabama experiments. The discordant findings were an enigma until after some years the experimental monkeys died. In each case, one of its own kidneys had not been removed and was still working contrary to the published experimental protocol. The grafted kidneys had been destroyed by rejection (The Scientist, July 2009).

These examples of cheating in science are damaging because they undermine public trust in the scientific discipline. For people to live together and function in a civilized manner, it is essential for individuals to have some degree of trust and affection between each other. Science as a verifiable truth normally helps to instill this.

Without trust, there can be no delegation of responsibility with the expectation that agreements will be fulfilled. Faith and trust are essential for many routine pursuits. It is necessary to be able to trust a friend to take care of one's baby. To fly with peace of mind, it is necessary to have faith in the workmanship of the plane's construction, and trust that due care will be taken of the complicated procedures required for the flight to take to the air and descend safely at the correct destination. The three case studies of cheating considered in the next chapter are some of the most harmful scandals which have undermined public trust in modern science.

Chapter 6
Three Cases of Corruption Involving Tissue and Cell Culture

The three scandals considered in this chapter were selected because of the enormous shock they caused, the widespread publicity that they brought to the scientific community, but also for the specific harm that befell those involved. They each led to a general denigration of science by the media. They are extremely sad stories and although they are not unique, they are fortunately not common.

In each of these three cases, I have tried to investigate how the idea to deceive germinated, how the fake science was presented and how the fraud was unmasked. In addition to these considerations, I will also examine who was dragged into the web of lies, how they reacted and what the outcome was. Finally, I will look at how the scandal was handled by the scientific community as well as by science journalists and the general media.

6.1 Case Study 1: William Summerlin

Origins

William Summerlin was born in South Carolina and he was educated at Emory University in Atlanta where he was once accused of cheating in an exam. He trained as a dermatologist and obtained a research fellowship at Stanford University in California where he became interested in skin transplantation. He had scientific ambitions and joined the internationally famous Immunology Department of Dr Robert Good in Minneapolis. Summerlin impressed Dr Good with his enthusiasm and strong work ethic. So much so, that when Good was head-hunted to become director of the world famous Memorial Sloan Kettering Cancer Institute in New York in 1973, he brought Summerlin with him and promoted him to the equivalent of full professor with extensive laboratory facilities and supporting staff.

Summerlin, at this point 35 years old, was popular due to his friendly charm and his devotion to his research. However, despite this, he also developed a reputation for sloppy record keeping and having a generally untidy laboratory. Although he was married with three young children, Summerlin often spent two nights a week sleeping in his laboratory to engage in his demanding "hands on" experiments. I had first heard about Summerlin a year previously when I was at the meeting of the Transplantation Society in San Francisco in 1972.

Summerlin had reported a special culture method for skin that after a few weeks supposedly rendered the tissue privileged immunologically. This meant that when grafted to animals and man the cultured grafts were not rejected. After presenting his findings to the Transplantation Society, Summerlin received a standing ovation and was besieged by some of the most famous scientists from all over the world, who wished to congratulate him and seek the secrets of his results.

Having in the past suffered many disappointments in the outcome of some of my own experiments, I was amazed that Summerlin claimed nearly 100% success, a very unusual finding with biological experiments. No mechanism was suggested by Summerlin to explain how apparently a simple culture technique for a few days could persuade complicated skin tissue to lose or hide their intrinsic Major Histocompatibility Complex, (the group of genes that code cell proteins so that the immune system can recognise and reject foreign tissue.)

Laboratories trying to repeat Summerlin's experiments wasted years of fruitless work. They initially blamed themselves for not doing the experiments properly, but the detailed methodology that should have been published by Summerlin did not exist. Eventually, Summerlin was suspended, when evidence emerged that he had faked the appearance of what he claimed to be successful skin grafts by painting them with black ink from a felt-tipped pen. The dastardly act was allegedly witnessed by a technician travelling in an elevator with Summerlin on his way to demonstrate the success of the skin grafts to his chief Robert Good.

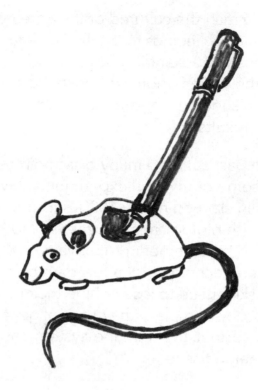

Fig. 6.1. A sketch of a skin graft on a mouse being coloured in with a felt-tip pen (RYC).

A book by the journalist Joseph Hixon *The Patchwork Mouse* published in 1976 gives a full account of the extraordinary Summerlin fraud. Despite many serious doubts being expressed by some of those involved in his laboratory, it took a long time to show convincingly that Summerlin was faking his results. This was largely due to the continuing loyal support by Robert Good, who was sucked into the scandal.

I had the privilege of visiting Robert Good in 1961, when he was working in Minneapolis. He was a very gracious host and devoted time to me, a young unknown surgical researcher.

His office was minute, situated between two busy laboratories. He explained that he deliberately chose a small office so that he would spend most of his time working in the laboratory and not be distracted by burdensome administration. When he was recruited by the Sloan-Kettering he became director of a huge scientific institute with responsibility for more than 100 staff, one of whom was William Summerlin. Good allowed his name to be amongst the authors of Summerlin's papers, without having participated actively in the experiments reported.

Robert Good became inadvertently a victim of the principle identified by Professor Laurence Johnson Peter (later explored in detail in sub-chapter 9.3.) in which scientists, burdened by the bureaucratic demands of a new promotion are unable to adequately supervise their laboratories. Good later admitted that the demands, responsibilities and busyness of his director's post, "prevented him from providing day-to-day supervision which viewed now in retrospect, might have prevented or terminated more quickly this regrettable affair".

Peter Medawar, the acknowledged world leader of transplantation immunology, also wrote a book on the Summerlin case, *The Spotted Mouse*. In this, Medawar speculated that in his early experiments, Summerlin observed skin grafts that may have been successful, because they were being performed between strains of mice which were mistakenly very similar to each other. This would have permitted graft acceptance. However, either failing to recognize or accept this error, Summerlin continued to maintain his mistaken interpretation and began to fall off the cliff face of credibility, digging a hole of lies and deceit for himself and others to fall into.

At a subsequent investigation, it was revealed that Summerlin had also faked the appearance of successful cornea transplants. He did this in rabbits by grafting an untreated human cornea on one of the rabbit's eyes which was totally rejected and fully opaque whilst claiming that the opposite eye, which appeared normal, had been grafted with a human cornea after culture apparently using his method. Evidence emerged that the crystal-clear eye had not been grafted at all!

According to the President of the Sloan-Kettering, Dr. Lewis Thomas, the disgraced Summerlin was said to have been suffering from "an emotional disturbance of such a nature that he has not been fully responsible for the actions he has taken or the representations he has made". Perhaps as a result of the Summerlin case, Robert Good was not rewarded with the Nobel Prize that many of his colleagues had expected him to receive, for his important studies on the function of the thymus gland. The Sloan-Kettering Institute, as well as science in general, suffered severely as a result of the scandal.

6.2 Case Study 2: Haruko Obokata

Origins

At school in Japan, Haruko Obokata was an outstandingly diligent and intelligent pupil, from an early age she held the ambition to become a scientist. She graduated in Applied Chemistry from Waseda University in Tokyo and was enrolled as a PhD student. Following this, Obokata was accepted to work at the Harvard Medical School Department of Surgery, (where I had spent nearly two years in 1960-61). Obokata worked under the mentorship of an academic anaesthetist, Charles Vacanti, who personally had raised funds to support her scientific fellowship. Vacanti, who was not an experienced cell biologist, had been working for some years on a theory that adult cells, if stressed in a special manner might de-differentiate into pluripotent cells similar to embryonic stem cells.

Obokata was knowledgeable and dedicated to hard work, she had a brilliant curriculum vitae with experimental expertise in cell biology. She seemed to be an ideal choice to work on and develop Vacanti's ideas. Vacanti found her to be an outstanding laboratory scientist. He wrote "I tell everyone that you are the most intelligent, hardest working, nicest, most creative and driven scientist I have ever had working in my lab, also the most beautiful". This praise was exceptional from the head of her department.

She was popular with most of her colleagues in the lab and all were impressed with her enthusiasm for hard work

Haruko Obokata © japantimes.co.jp

Fig. 6.2.   Haruko Obokata in the lab (IPD).

maintained over long hours. A few, perhaps with green eyes, nicknamed her "The Little Princess."

In Boston, Haruko Obokata claimed to have developed a cell culture technique that could de-differentiate neonatal spleen cells into cells with the pluripotent characteristics of embryonic stem cells, later called Stimulus Triggered Acquisition of Pluripotency ("STAP" cells). She achieved this by using a slightly acid culture medium. Obokata's findings at first seemed to supersede the discovery of induced Pluripotent Stem Cells (iPS cells) by Shinya Yamanaka, which required a much more difficult and complicated process.

Two papers in which Obokata was the first author were published simultaneously in the January 2014 edition of *Nature*. This must be an almost unprecedented achievement for such a prestigious journal to publish two papers together by a young first author. The media began a frenzy with newspaper headlines celebrating how a few drops of vinegar could forge a breakthrough in biological science.

When Obokata returned to Japan, she was treated as a celebrity and a national heroine. Sadly, this adulation only lasted a few weeks until a number of independent laboratories reported that they we unable to repeat the STAP Cell experiments.

Following these reports, the celebrity veneer was quickly stripped off and Obokata was attacked viciously by the media. On one occasion she was even pursued into the ladies room of a hotel by reporters from the Japanese National broadcaster NFK. This fulfilled the unsavoury reputation of the media for creating idols with hype and then tearing them down with even greater enthusiasm. Obokata was rapidly deserted by her co-workers as if fleeing a sinking ship. Just previously, these same individuals had enjoyed being associated with her work and publications.

Yoshiki Sasai, a co-author of her disputed publications and a much loved and respected senior scientist at the Riken Institute in Kobe, committed suicide by hanging himself in the laboratory. He left a note for Obokata urging her to confirm the production of STAP Cells with further experiments.

Two extensive investigations by the Japanese authorities were unable to provide full clarification of how the scandal evolved. The second investigation did discover that some of the cells used in the experiments were real embryonic stem cells, but derived from a different mouse strain to those specified in Obokata's publications. The investigation's report ended with the comment that the wrongly identified cells probably did not contaminate the culture by accident!

In 2016 Obokata published a book to give her side of the story, "Ano Hi" (That Day). In this she claims that she was framed by Teruhiko Wakayama, a senior associate, who had deliberately contaminated her cell cultures and that she regrets not being able to deliver a public apology.

Teruhiko Wakayama was a senior colleague of Obokata at the Riken Institute. Initially, he was an enthusiastic collaborator in the STAP cell work. He was senior author in one of the Nature papers and co-author in the other.

He left the Riken Institute and continued to support Obokata; providing cells in culture for her experiments, whose provenance was later disputed. Obokata subsequently accused him of having been involved in her downfall by framing her as the guilty scientist in the scandal. She alleged that the cells that she had used in her experiments were provided by Wakayama who had since moved to Yamanashi University. There has been no explanation of a motive for such an act.

The Riken condemned the fraud, but did give Obokata the chance to try to achieve redemption by repeating her experiments under very closely monitored conditions. She found the continuous surveillance intrusive and disturbing. After months of endeavour, she was unable to produce STAP Cells.

The Riken's final report found that Obokata had falsified and fabricated data. The so-called STAP cells were genuine embryonic stem cells from three mouse strains different to those described in her publications. Moreover, they also did not think that the contamination was accidental.

The report criticised Wakayama who was head of the laboratory where Obokata worked, and Yoshiki Sasai who was a major figure in compiling the final versions of the STAP cell publications. They were admonished for their lack of appropriate scrutiny of the data, figures and laboratory protocols in Obokata's work. In one year, the whole STAP story had unravelled from jubilation to disgrace.

Currently there is no more information on the background and evolution of this tragic and mysterious story. It is not clear what went wrong. Obokata's mentor at Harvard Medical School, Charles Vacanti retired at age 65 as an Emeritus professor. He supported the withdrawal of the two papers but still believed that stressing cells in tissue culture could result in conversion of some cells to pluripotent stem cells.

There has also been no report of any official internal or independent investigation carried out at Harvard Medical School.

Of the three corrupt cell culture scandals considered in this book this is the most extraordinary. It evolved as a complex and mysterious detective story with apparently devious activity. In the course of a short time, a young scientist became internationally famous, puffed up like a magical balloon into a national celebrity by an enthusiastic media. Eventually, the whole edifice collapsed and the glittering stars of the heroine rapidly turned to dust, with vilification from the same media that had erected the paragon.

Many important features of this case remain unanswered, who contaminated Obokata's cultures? What was the motive? No one would benefit from the unmasking of the fraud unless it was an act of revenge. The death by his own hand of a much loved and respected senior scientist, who persisted in believing that the fake work was genuine, added a tragic chapter to the sombre saga. Because the evidence is incomplete and insufficient to determine the whole truth, the jury remains out and may never reach agreement.

My belief is that more competent supervision of Obokata's work and an improved journal review policy might have prevented the tragedy.

Further Reading:

For more information on this case John Rasko and Carl Power covered the story in *The Guardian*, 18 February 2015. There is also an excellent essay on this case published in the New Yorker Annals of Science by Dana Goodyear, 29 February 2016.

6.3 Case Study 3: Paolo Macchiarini

Origins

Paolo Macchiarini was born in Basle, he qualified in Medicine at Pisa University in Italy and trained as a surgeon. He worked in a number of hospitals in Europe and became interested in the trachea as an organ that by transplantation or substitution might provide therapy for patients who had lost function in part or the whole of their tracheas. He moved swiftly to clinical application, without at any period of his training being exposed to rigorous surgical science in a recognised surgical or biological experimental laboratory. He frequently moved institutions between 1999 and 2017, never settling for more than a few years in any one place. During this time he lived in Paris, Hanover, Barcelona, Florence, Stockholm and Kazan in Russia. He also acquired various honorary appointments, including two in the UK in Bristol and London.

From about 2011 onwards, Macchiarini appeared to believe that bone marrow mesenchymal stromal cells could be seeded on rather crude plastic tubes and that these cells would differentiate to produce a functional replacement for an absent or damaged trachea.

The mammalian trachea is a complex vital organ with an abundant blood supply and a highly specialised lining of epithelial cells. These cells are equipped with minute and dynamic moving hair-like structures called cilia that sweep the contents of the lumen of the trachea upward to be removed. The process is facilitated by a continuous stream of mucous secreted by goblet cells in mini-organs that are part of the structure of the tracheal wall (Fig. 6.3).

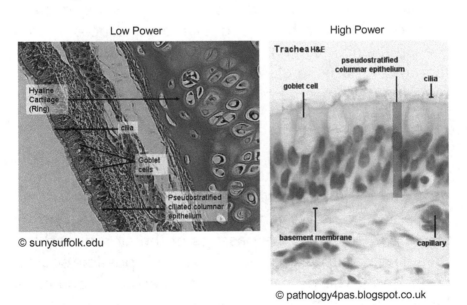

Low Power   High Power

© sunysuffolk.edu

© pathology4pas.blogspot.co.uk

Fig. 6.3. Collage of histology of normal trachea low and high power showing complex tissue.

A vital component of the tracheal anatomy is a well developed vascular system of arteries, capillaries and veins. The scaffold consists of living specialised cartilage that provides a degree of rigidity combined with some elasticity, mimicked by the flexible air tubing in anaesthetic and other breathing tubes.

To expect mesenchymal stromal cells to adopt these multiple and quite separate tasks following seeding on a simple plastic tube is asking a lot. Indeed, attempts to do this had been repeatedly tried by others in animals for decades with universal failure. It is not surprising that neither Macchiarini nor other experimentalists have so far been unable to publish any convincing evidence that this could be done. The best that might be hoped for was for some of the recipient's own genuine tracheal cells to creep over the plastic for short distances. In this scenario, function might only be expected for very short lengths of plastic implants and certainly one would not expect a good result from a full-length implant of six to nine inches.

In addition to this, the air that passes through the tracheal tube is exposed to the atmosphere which contains potentially pathogenic bacteria and fungi; unlike for example hip replacements, which are completely embedded in the body. Therefore, the passage of the air required for breathing was itself an added hazard as a possible source of infection. The formation of pathogenic bacteria-contaminated biofilm on the tracheal surface was almost inevitable.

Macchiarini was appointed to the staff of the famous Swedish Karolinska institute, some of whose academic staff have the responsibility of awarding the Nobel Prize in Medicine and Physiology through the independent Nobel Assembly.

Extraordinarily, Macchiarini's appointment was not preceded by the generally accepted scrutiny of his curriculum vitae required for all academic posts in all universities worldwide.

Despite there being no record of large-scale animal experiments being performed, Macchiarini began treating patients with this untested procedure. I attended a meeting in London at which Macchiarini presented his results of complete tracheal replacements in patients with excellent outcomes.

Unfortunately, following Maccharini's report of his tracheal replacement cases doing very well and a publication in the *Lancet*, there were worrying observations made by physicians looking after his patients. Some of them had suffered from grievous infective complications and had died. Nevertheless, an initial internal investigation exonerated Macchiarini.

Because of the severe pain, prolonged suffering and distressing deaths reported by the doctors looking after Macchiarini's patients, there has been extensive scrutiny of the whole tragic series of events. The course of events that followed was complicated as many individuals were involved in different capacities.

An external review by Professor Bengt Gerdin found scientific misconduct in all six published journal articles relating to Macchiarini's work. Despite this, the Karolinska Institute chose to clear Macchiarini and his co-workers. Furthermore, the Director of the clinic at the Karolinska Hospital, Anders Hamsten, threatened to sack the whistle-blowers.

In January 2016, Bosse Lindquist, a respected and experienced director of documentaries for Swedish National television, produced a three-part detailed investigation of Macchiarini's tracheal operations called 'The Experiments'. It was very disturbing to watch as some of the patients selected for the subsequently disastrous procedure seemed to be managing quite well with standard tracheostomies and what appeared to be perfectly safe lung ventilation. The institutions in Sweden and Russia, where the operations took place, appeared to have little control over what was going on.

There were reports of further deaths and the Prime Minister of Sweden, Stefan Löfven, on a visit to Mumbai, called the case a scandal and stated that "it was very important for a further investigation to be carried out to determine what had gone wrong and correct what needed to be corrected" (*Expressen* 13 February 2016).

Following this, a further independent inquiry was requested by the Karolinska Institute. This was carried out by the expert group on scientific misconduct from the Central Ethical Review Board (CEPN). The expert group appointed two external experts, Professor Martin

Björck, Uppsala and Professor Detlev Ganten, Berlin. They similarly to Bengt Gerdin found that "there was scientific misconduct in the articles in question."

Following this second enquiry, the head of the Karolinska Institute, the secretary of the Nobel Prize Committee for Medicine and Physiology, and others implicated in the scandal resigned or were dismissed. The Swedish government sacked all members of the Karolinska Institute's board who remained in position and an investigation into Macchiarini was launched by the Swedish Prosecution Service.

Bo Risberg, Professor Emeritus of Surgery at the University of Gothenberg and a former chairman of the Swedish Ethics Council, called for the Nobel Prize to be suspended for two years as an "apology" to Macchiarini's patients and their families. He said that the events amounted to the biggest research scandal Sweden has experienced in modern times. Risberg would later note that "it is very strange that it should take a TV programme to make this public... everything was swept under the carpet." The failure to do pre-clinical tests on animals, he added, was "the worst crime you can commit".

In developing new and radical surgical procedures there is a long established and accepted ethical route to be followed, in which extensive animal investigations are mandatory to show as far as is possible, that the chances of a good result in patients should be expected. No such thorough experiments had been done by Macchiarini or others. Furthermore, as described above, there was no precedent in biological research to suggest this procedure could be successful. For bone marrow derived mesenchymal stromal cells to change into the many different specialised cells that would be required to fashion a new, normally functioning trachea on a plastic scaffold was always an uphill and likely impossible task.

As in the previous two scandals described in this book, the head of a worldwide famous institution was sucked into the scandal. The director of the Karolinska Institute defended Macchiarini forcefully. The "whistle blowers" looking after the dying patients had alerted the authorities to the malpractices that they had observed and were not listened to; in fact, they were personally rebuked.

During the course of the proceedings, Dr Pierre Delaere, Professor of ENT in Leuven Belgium and a head and neck surgeon, offered severe criticism of the disastrous surgical procedures that had taken place. He described the operations as based on "one of the biggest lies in medical history, because Macchiarini was doing something that is impossible from a theoretical point of view." He further volunteered that he would rather face a firing squad for a speedy painless death than submit to the prolonged suffering, pain and distress that would follow being operated on by Macchiarini (*The Guardian* 9 October 2017).

After an investigation, the Central Ethical Review Board published its findings on 27 October 2017 concluding as follows:

> *The Expert Group state that the transplantations are described successfully in the articles, which*

is not the fact. The Expert Group also establish that the information in the articles are misleading and beautifying regarding the patients conditions and furthermore that information has been withheld in this purpose and that this constitutes scientific misconduct. In addition, there is false information of ethical approval, which also constitutes scientific misconduct.

The Expert Group finds that all co-authors to the six articles are guilty of scientific misconduct. The main responsibility lies on Paolo Macchiarini as the main author and research-leader and others who have had a more prominent role in the research and authorship. The more detailed responsibility and the future consequences for the respective authors is up to the employers to decide.

The Expert Group request the different publications to withdraw all six articles.

A Romantic Deception

Macchiarini had great charm and was exceedingly convincing in his deceit. In fact, in his private life he was able to maintain a deception to a degree that is difficult to believe. Coincidental with the medical disaster, a romantic betrayal of melodramatic dimensions developed between Macchiarini and Benita Alexander, an NBC film producer from New York. Alexander had been charged with overseeing a programme on Macchiarini and was swept off her feet by the charm of the handsome Italian surgeon. She fell deeply in love with him, ignoring his many serious character defects as her reasoning was impaired by her infatuation.

Macchiarini promised to marry her with extra icing on the wedding cake. He told her that he had secured an agreement with the Vatican that the Pope himself would officiate at the ceremony. Mr Putin and President Obama among other notables would be invited to the wedding. However, a friend of the betrothed discovered that the Pope would in fact be in South America on the day appointed for the wedding. A visit by Alexander to one of Macchiarini's residences in Barcelona provided an unexpected and unwelcome meeting with Macchiarini's actual partner as well as a later discovery that he was still married to his wife in Italy.

When this web of lies was exposed, the would-be-bride reacted in the manner suggested by Congreve's proverb:

© Benita Alexander, Private Collection

Fig. 6.4. Trip in Venice — Vanity Fair, 5 January. 2016. (Permission generously given by Benita Alexander from her private collection).

'Heaven has no rage like love to hatred turned, nor hell a fury, like a woman scorned. (William Congreve, The Mourning Bride, 1697). Alexander found revenge for being deceived, by publishing the whole story with suitable illustrations in the magazine *Vanity Fair*, in January 2016.

Of the three scandals considered in this book, Macchiarini caused by far the most serious suffering. He deceived patients who put their trust in him. In all three cases it is alarming how easily senior scientists were sucked into the scandals. Each case led to a denigration of their institutions as well as to a denigration of science more broadly. They each highlight that whilst science should be a verifiable truth, scientists themselves are not immune to the fraudulent behaviour which is instinctive in human nature.

Chapter 7
Common Characteristics of Scientific Fraudsters

7.1 Without Conscience and the Walter Mitty Syndrome

In the course of writing this book, I have become intrigued by the distorted thought processes that go through the minds of science cheats and whether it is possible to identify common themes between them. In this chapter I will examine a few of the common characteristics of scientific fraudsters, tracing how minor deceptive traits in their personalities can germinate into strangulating bindweeds.

The Canadian psychologist, Robert Hare, in his book *Without Conscience* (1999), outlined a personality trait in some psychopaths involved in crime. They are self-centred and may appear to be charming in how they manipulate others for their own personal gain. They are often unable to empathize with other people and lie frequently, without demonstrating any evidence of having a moral conscience. They do not respond to treatment and are incurable. These features seem to fit some of the scientists who are responsible for major falsification in their work.

In James Thurber's story of the Secret Life of Walter Mitty the meek mild-mannered protagonist, imagines a vivid fantasy world for himself in which he adopts different superhero personas, whilst he is taking his wife to a hairdressing appointment.

He is first a fearless wartime fighter pilot flying through a terrible storm and involved in desperate dogfights with Messerschmitts in the Battle of Britain. He then becomes a brilliant surgeon who visits an operation in progress, where the local team are failing and the patient's life is rapidly fading away. Walter Mitty steps in and with extraordinary dexterity stops the fatal haemorrhage, saving the patient. Following this, he becomes a prominent member of a criminal gang and is prosecuted as a murderer. His eventual fate is to be executed by a firing squad, before which he scorns the blindfold offered to him, and faces the guns with a defiant smile.

There seem to be certain features of Walter Mitty's character evident in the science fakers that I have considered in this book. Their desire to be successful, especially if given false hope by an initially positive first result (even if this is erroneous), can lead them to experience or imagine a life of fame and success which is difficult to give up if their experiments later turn out to fail. Similarly, although most of them are intelligent, they all seem to be seriously lacking in common sense.

7.2 Cloud Cuckoo Land

In the 414 BC play *The Birds by* Aristophanes, the plot concerns two friends accompanied by their very intelligent corvid pets, a crow and a jackdaw. They discuss how fed up they are with the politics and widespread corruption in ancient Athens. They turn to their avian companions and suggest that if all the birds get together they could build a much better city than Athens above the clouds. The birds collaborate to construct a wonderful city in the sky called *Cloud Cuckoo Land*. Cloud Cuckoo Land turned out to anger the Gods and it develops similar defects to Athens itself. The deceptive illusion of Cloud Cuckoo Land meant that it ultimately became identified as a dwelling place for crazy people.

One of the most prominent features that the scientific fraudsters considered in this book appear to share is a disconnection from the real world. Their imagined successful personas and their temptations to ignore negative results means that they might be best described as *Walter Mittys*, who have set up house in Cloud Cuckoo Land! *(Fig. 7.1.)*

Cloud Cuckoo
Land
"The Birds"
Aristophanes
© British Museum

Laconian
Calyx 540
B.C.

Fig. 7.1. Image from ancient Greek ceramic pot dated 500BC, British Museum (IPD).

Chapter 8
Originality and Genius

8.1 Genius and the Pursuit of Originality

The motives of scientific fraudsters are complicated and vary. An ambition to be a famous celebrity and admired by scientific colleagues and even the general public as a genius would appear to be a feature of the distorted thought processes of some dishonest scientists. In order to achieve this status, there is often a strong desire amongst scientists to develop ideas which are considered original. This desire can become overconsuming and spiral out of control leading some scientists towards fraudulent behaviour.

Originality suggests new and independent ideas which are different from anything anybody else has previously conceived. However, in science, most advances have to varying degrees, been derived from work already done by others and it is usually impossible to divorce completely new work from what has gone before.

Isaac Newton acknowledged the legacy of previous observations when he remarked that if he could see more clearly than others before him, it was, "by standing on the shoulders of giants". Nevertheless, Newton was involved in a long and unpleasant dispute with Gottfried Leibniz on the priority of who had invented the mathematical calculus or whether they had both arrived at the idea independently. Newton also had bitter quarrels over claims of priority in the discovery of gravity and the nature and wave theory of light with his contemporary, Robert Hooke, then president of the newly established Royal Society. When Hooke died, Newton succeeded him as president and in the course of the Royal Society moving to new premises, the only portrait painted in life of Hooke disappeared and has never been found! Robert Hooke was certainly a giant in science, but it is doubtful if Hooke's shoulders were on Newton's mind when he acknowledged the work of his scientific predecessors that influenced him so profoundly.

Perhaps the discovery of the circulation of the blood by William Harvey can be accepted as a completely original idea. In 1628, he postulated that there had to be a connection between the arteries and veins, although they were too small to see. This hypothesis was based on logical deductions and was contrary to the accepted dogma of the teachings of the Greek physician and surgeon, Galen, who worked at the time of the Roman Empire. It was many years later that these vessels, which are called capillaries, were viewed directly. This was achieved by Marcello Malpighi in 1661 when using a microscope to observe frog lungs.

Examples such as Harvey are rare and the pursuit of originality, especially when it defies perceived scientific wisdom, can be a factor which leads scientists to deceive.

8.2 Originality in the Arts

Humankind's desire for originality is similarly manifest in the world of art. Before the invention of photography, the skill and ability to depict images and sculptures convincingly were the attributes most appreciated in the assessment of art. In the 5th century BC, in ancient Greece, a competition between the two greatest artists of the time was held to determine who could depict most accurately a three-dimensional subject in a painting. The first competitor, Zeuxis unveiled his image of a bowl of fruit. It was said to be so beautiful that a bird flew down and tried to eat a grape from the painting.

When the second contestant, Parrhasius, was asked to unveil his painting he hesitated and said it could not be done. The audience thought he was overwhelmed by the competition, but then it dawned on them that the curtain itself was his painting. There was a general agreement that Zeuxis had deceived the bird, but Parrhasius had deceived his fellow painter and the audience, so he was awarded the prize.

Throughout the arts, originality is continually lauded and lack of originality is often dismissed by the experts as worthless, but as with science, it can be difficult or even impossible to exclude the influence of the work of those who had gone before. There is an easily traceable connection between the music of Haydn, Mozart and Beethoven, yet each of these composers is universally regarded as brilliant and original. No one else could have composed their unique music, often with unmistakable signature features recognized by other musicians after hearing just a few bars.

Like many scientists and the aforementioned musicians, the great painters of the Renaissance were intimately aware of the work of each other, yet there was little evidence of friendship and collaboration between them. Rivalry was more common, for example, between Leonardo da Vinci, Michelangelo and Raphael. They each knew exactly what the other artists were doing in their individual struggles to be the best. Furthermore, each achieved lasting greatness with unique originality that cannot be brushed aside despite obvious derivative features acquired from other artists. The same is true of the Impressionist painters who influenced each other and also often obtained inspiration from eastern Japanese and Chinese artists. The walls of Claude Monet's house in Giverny are bedecked by many beautiful Japanese woodblock prints.

In science, like art, true originality is very rare and should be treated with healthy scepticism. My scientific hero and friend Peter Medawar, if told of scientific work that seemed too good to be true would reply, "Curious! If it can be repeated in another laboratory it would be interesting." This is an attitude which should always be present in scientific research.

That being said, originality in science is not entirely of the same nature as originality in art. The genius of Rembrandt in both the intensity and beauty of his masterpieces could never have been executed by anyone else, unless copied with extraordinary skill by a fraudster, or a work produced by an artist painting together with the master in his studio.

Fig. 8.1.   Portrait of Sir Peter Medawar (RYC).

Rembrandt probably worked in a very busy studio with several talented apprentices. As the master he would view work in progress and add or change some features in his pupils' paintings. He may have asked one of his apprentices to paint part of his own painting not yet finished, for example fruit or flowers.

In attributions of authenticity years later it can therefore be difficult to tell exactly how original any given painting is. This is particularly true when a genuine masterwork has been severely damaged by the ravages of time. Intense restorative efforts by many hands may modify and corrupt the original so greatly, that doubts can be raised as to how many of the original brush strokes were made by the artist themself.

When special committees are set up to select genuine Rembrandt paintings from copies or fakes, they base their decisions not only on forensic evidence, but also on their assessment of the quality of the work. In a recent committee they dismissed the possibility that like other mortals, Rembrandt might occasionally have a bad day at the office.

Originality in science is different because new advances cannot rely on the prodigious skill of one individual. New discoveries must be repeatable and published with an accompanying scientific method so that experiments can be replicated in other laboratories. Nevertheless, the desire to be original can both motivate scientists to cheat and should also be regarded as a red flag for journals when reviewing supposedly groundbreaking advances.

Chapter 9
Problems Facing the Discipline of Science

The cases of cheating considered in this book highlight some of the most problematic issues facing the discipline of science more generally. In this chapter I will discuss some of these problems, examining how they hinder the progression of science and help allow deceptive practices to prosper.

9.1 The Two Cultures

In 1959, the academic physical chemist, novelist and philosophical polymath C. P. Snow gave a Rede Lecture in Cambridge published as "The Two Cultures and the Scientific Revolution". In this, Snow argued that the intellectual life of the whole of Western society was split into two cultures, namely the sciences and the humanities. This was a major hindrance in solving the world's problems. As advances in science accelerate, the division persists and the importance of science to society continues to have little influence on national policies. Despite two of our recent Prime Ministers in the UK having held degrees in science (Thatcher — Chemistry, May — Geography), scientists have constitued a small minority in representative government.

This problem is manifest in even the most educated echelons of society. As Snow stated elegantly:

> "A good many times I have been present at gatherings of people who, by the standards of the traditional culture, are thought highly educated and who have with considerable gusto been expressing their incredulity at the illiteracy of scientists. Once or twice, I have been provoked and have asked the company how many of them could describe the Second Law of Thermodynamics. (The Second Law of Thermodynamics can be difficult to recall and understand by many scientists!) The response was cold: it was also negative. Yet I was asking something which is the scientific equivalent of: Have you read a work of Shakespeare's?
>   I now believe that if I had asked an even simpler question — such as, what do you mean by mass, or acceleration, which is the scientific equivalent of saying, can you read? — Not more than one in ten of the highly educated would have felt that I was speaking the same language. So, the great edifice of modern physics goes up, and the majority of the cleverest people in the western world have about as much insight into it as their Neolithic ancestors would have had".

This level of scientific illiteracy is problematic and hinders wider awareness and scrutiny of scientific advances in society more broadly.

9.2 The Scientific Tower of Babel

When the Royal Society was founded in the early 17th century, the original members came from a variety of backgrounds. They met in London coffee houses and taverns, and the new society provided them with a forum in which dissertations could be understood by all; varying from natural history, chemistry, cosmology, mathematics, philosophy and virtually any subject that was thought might be of interest to educated gentlemen.

The proliferation and separation of the sciences in recent times has had what could be called a Tower of Babel effect. According to Genesis, the purpose of the attempted construction of the Tower of Babel was to reach Heaven. However, God, realising this plan, determined to confound the voices of those building the tower so that they could not understand each other, and the tower was never completed.

The lack of mutual understanding between different scientific fields is very worrying as modern science often requires collaborative research between experts from different disciplines. Although these experts may be able to communicate in general matters of mutual goals and strategies, when collaboration requires contributions from differently trained scientists, each with his or her own knowledge and skills, serious difficulties can arise.

The individuals schooled in completely different disciplines and ways of thinking, cannot be expected to understand the intricacies and potential pitfalls within which scientists from other separate special faculties would be very familiar.

The Tower of Babel effect is something I observed myself when working in a collaboration which required both molecular biological and surgical expertise. Not only were there complete differences in the training, knowledge and understanding between the experimenters, but communication was also not easy due to the two different native languages involved. These defects in understanding can leave cracks in the collaborative efforts of research plans that are attractive pastures for fakers to exploit.

9.3 The Peter Principle

Compounding the issues which can be created by the Two Cultures and the Tower of Babel effect, the system of promotion for scientists and the increased bureaucratic demands that promotions typically entail has frequently hampered many of the brightest individuals in given scientific fields. This general and often disastrous pattern of promotion was explained by Professor Laurence Johnston Peter in 1968.

In almost all large institutions and especially in hierarchical academic, political and commercial offices, there is a common pattern of promotional policies. An individual identified as productive and hardworking will be offered a new post with more work, responsibility, a higher salary, a bigger office and more secretaries. If the job is done efficiently without complaint, an offer is soon made for further promotion, another step up the hierarchal ladder. Eventually the individual is overwhelmed by work and there is no longer time to fulfil all the tasks that are required. Corners are cut and a realisation that a level of incompetence has been reached can be difficult to accept and is normally anathema to the high flyer.

The requirements for success in the new job can be quite different from those that were identified as admirable further down the ladder. An excellent and innovative engineer may not be possessed with the social and diplomatic skills, charm and ability to enthuse and encourage others, negotiate delicate deals and accept compromises. An abrasive but "can do" personality may not be suited for what is often a totally bureaucratic role at the top. Where to draw the line on delegation is commonly misjudged and obsessive micro-management might be interpreted by junior staff, often correctly, as a lack of trust. On the other hand, a laissez faire approach can lead to a lack of essential supervision which can be the perfect setting for fraudulent scientists to prosper.

9.4 The Prize Problem

Another problem which can hinder the honest progression of science is the set of criteria which determine how prizes are awarded. It is not surprising that when a new discovery is made, those involved will be anxious to establish and safeguard the priority and recognition of what they have done in the advancement of science. Failure to do this may result in their contribution being bypassed or appropriated by somebody else. However, in the past 50 years, more and more scientific work is conducted collaboratively by members of a group.

Consequently, it can be difficult to disentangle individual contributions regarding who said what, and in which order. It is a common experience for new ideas and suggestions to emerge from discussions at morning coffee, afternoon tea or in a pub. This can lead to controversy that is not always amicable and may be bitterly disputed for years after, if attribution of a discovery was felt to be incorrect.

The credit for originality can result in the award of a much sought-after and prestigious prize that may be shackled to archaic rules laid down by the benefactor many years previously. The rules and their interpretation may have in some cases moved away from the original wishes of the deceased donor of the prize. For example, the greatly esteemed Nobel Prize, with its rule of not being allowed to award more than three recipients for a single contribution, might be considered no longer fit for purpose given that work is often collaborative and may involve many more than three individuals or laboratories in a joint project. The current culture of scientific practice with widespread collaborative research, often involving a variety of separate institutions is fundamentally different to that pertaining when Nobel wrote his will.

Jocelyn Burnell, as a postgraduate student, was the first to discover radio pulsars, however, it was her mentor and the head of the department who were awarded the Nobel Prize, whilst she missed out. It is no wonder that young scientists can be very anxious as to what may happen when they believe that they have made a major discovery. The potential for winning prizes can lead scientists to be secretive and non-transparent about their work. This behaviour can quickly spiral out of control and develop into lies and deception.

For science to enhance its current role in society, it must endeavour to confront some of the challenges explored in this chapter. By expanding participation and collaboration in and between the sciences, whilst also addressing the problems relating to promotions and prizes, the field of science will both increase scrutiny on new discoveries and reduce the possibility for fake and corrupt practices to prosper.

Chapter 10
The Problems Facing Young Scientists

Having identified some of the problems facing science more broadly, in this chapter I will examine the issues specifically facing young scientists who are just starting out on their careers. Here I will also offer some advice for how young scientists might handle disappointments and manage expectations for their careers ahead.

10.1 Dealing with Disappointment

Most youngsters are attracted to study science due to a natural curiosity to discover new and wonderful things and a desire to help explain how the world works. Once launched into the actual practice of experimental routine, unless they are very lucky, they will probably be faced with disappointments.

Even if accepted by a famous professor, they may be sucked into the requirements of the overall strategy of the department. Within the institution, there may be a variety of laboratories in which the young scientist might be offered a place. Although the head of the department may have been involved in their appointment, the larger and more renowned the department is, the less likely the department head will be able to work in the lab.

It is not uncommon for young scientists to feel that they are being used as a cog in the wheel of a new technique or procedure discovered in the department, for which it has been recognised internationally. Alternatively, they may be assigned to perform an experiment which fails miserably.

The choice of mentor for new PhD students may be difficult and in some cases the student may not even have a chance to contribute to the selection. If, as is likely, the laboratory is involved in an ongoing project, the student will probably be directed to participate in the main research effort of the department with instructions on new techniques that will fit in to the mainstream of work. Individual autonomy for the student will be uncommon, with mundane and sometimes monotonous tasks more likely to be the student's responsibility.

10.2 Managing Expectations in Cell Biology

If the work of a laboratory involves stem cell cultures, the young scientist will have to learn the details and techniques required to care for the special cells selected. This will involve how to keep them alive, how to nourish them and if necessary how to differentiate them in the direction that is needed in the experimental design. The care of the cells can be similar to that required in very skilled and delicate gardening. It usually involves attention to detail and quite likely will necessitate visits to the laboratory at unsociable hours as well as during holiday periods. Trips may also be needed during the night to a cold and empty laboratory. These often solitary chores may be very far from what the student had expected and quite different from the picture they had envisaged of themselves doing good for humanity, solving important and difficult problems leading to recognition, prizes, fame and advancement in the scientific hierarchy.

If the cells under the care of the student seem to be behaving as hoped for this will bring delight and joy. Repetition of the experiment will be essential as well as checking that the origin of the cells is correct and the results have been observed critically with great care. If the experiment cannot be repeated this will be very disappointing. Questions have to be answered to verify the nature of the cells and to make sure they have not suffered in the culture procedure. For example, has the culture been contaminated by foreign cells or infection? It is not surprising that these matters can produce much soul searching, worry and temptation to handle the matter with bias and perhaps even stray into areas of dishonesty. It requires courage and fortitude to resist this temptation.

A life in science should be driven by a natural curiosity, and a dissatisfaction with current established explanations of problems that seem to be important. In view of the inevitable disappointments that are naturally depressing, it is essential to realise that lessons are to be learned from mistakes. A careful analysis of what went wrong should ensure that the same mistake is not repeated.

There is a good chance that a period of hard thinking and reflection will boost a leapfrog over the current hurdle into more encouraging territory. The concept of the experiment may rest on ideas that are later to be found incorrect. To come to grips with failed experiments, a science student needs a robust personality and if still convinced that the approach to the problem is on the correct track, then perhaps, a minor modification in the protocol may be rewarded with success.

Louis Pasteur identified in himself that, 'my strength lies solely in my tenacity'. Tenacity can be the most important characteristic to keep hold of the problem like a dog with a much-prized bone, in the faith that with more chewing eventually the bone will shatter and there will be delicious marrow inside.

10.3 Follow the Example of Frederick Sanger

Although there are many different ways of actually doing science, I think an excellent example was set by Frederick Sanger who was awarded two Nobel Prizes in the same field, an extremely rare distinction. Sanger attacked two separate well-defined problems. First the molecular structure of the vitally important protein insulin and second the order, arrangement and significance of DNA and RNA bases in the synthesis of proteins. The clarification of this latter thorny challenge directly helped in the mammoth task of unravelling the whole human genome.

Sanger worked in a modest laboratory, with minimal technical assistance. After his work on insulin, Sanger did not publish any further scientific studies for some time.

A fact-finding visitor from the Medical Research Council, that is responsible for the financial organisation of the Laboratory for Molecular Biology in Cambridge where Sanger worked, is said to have asked Max Perutz, director of the institute, "What is Dr Sanger doing now? He has not published any work for some time." To which Perutz replied, "Dr Sanger is thinking". Shortly after, Sanger was awarded his second Nobel Prize!

Sanger himself looking back over the years is quoted as saying, "Of the three main activities involved in scientific research, thinking, talking and doing, I much prefer the last and am probably best at it. I am all right at the thinking, but not much good at the talking." After his first major achievement, Sanger entered a period that he described as "lean years with no major success". He gave advice on how to deal with these periods that affect many careers, not just scientific ones:

"I think these periods occur in most people's research careers and can be depressing and sometimes lead to disillusion. I have found the best antidote is to keep looking ahead. When an experiment is a complete failure it is best not to spend too much time worrying about it but rather get on with planning and becoming involved in the next one. This is always exciting and you soon forget your troubles."

It has been suggested that pursuing scientific curiosity can have similarities to seeking a black cat in a dark room, when you are not even sure that the cat is in the room. That being said, cats have excellent vision in the dark and the lost cat may take a liking to you, spring into your arms and whisper the solution to the problem you have been trying to solve into your ear. A life in science is clearly not for the weak-hearted and persistent passion to become a scientist is an essential requirement.

I hope that these insights into the scientific philosophy of Frederick Sanger might be encouraging for young budding scientists. Managing expectations and anticipating disappointments will hopefully help individuals beginning a vocation in science to have a successful career and steer clear from any temptations to deceive.

Chapter 11

What Can be Done — A Few Suggestions as to How Cheating in Science Might be Curtailed

In this last section, which is based more on hope than expectation, I will make a number of tentative suggestions that could lessen the incidence and the temptation to cheat in science. These recommendations involve changes in the hierarchical structure of scientific laboratories, a greater scrutiny in the recruitment process and an improvement in the way that scientific articles are reviewed, published and distributed.

Underpinning these recommendations is a careful consideration of the human nature of scientists and particularly the emotional stresses which may arise from personal conflict and ambition.

11.1 Reconsidering the Hierarchical Structure

Because the way in which science is practised has changed over the years, especially in the organisation of institutes involved in 'Big Science', the hierarchical structure of scientific and medical institutions needs a rethink. Individual laboratories should have more independence and responsibility in decision making to avoid burdening the overworked head of the institute with micromanagement, that he or she cannot be expected to handle properly.

A transparent system of supervision and mentorship should be a standard set-up in all laboratories, in which the supervisor and the scientist being mentored are obliged to meet at regular intervals, which are required to be documented and agreed together by both parties, with a summary of what transpired and decisions made.

With the background outlined in the rest of this book and the suggestion that deceit is a hardwired human attribute, similar to that found in some animals, the possibility of removing this instinctive trait from science is a difficult task. However, differing from animals, humans make a conscious choice when they decide to lie and deceive.

Because of human conscience, doctors on qualification are required to follow the Hippocratic Oath, a few simple ethical rules based on common sense with an overall requirement not to harm the patient. In science, a course of scientific ethics, history and good practice should precede any experimental work. A similar oath might also be an expedient soft measure for young scientists, as a solemn declaration that he or she will not falsify or cheat in their scientific work.

11.2 The Journal Review Process

The unfolding of the STAP Cell disaster and scandal was a serious blow to the integrity of Science. I put forward a few suggestions that I felt might be of help in this letter sent to *Nature*.

> **Original Correspondence to *Nature* on the subject of delay and difficulties in getting work published (8 March 2016) An edited version of this letter was published by *Nature* (535: 493, 26 July 2016)**
>
> *The recent discussion on this subject in an insightful article in Nature (Kendall Powell, Nature 16 February 2016) was helpful in explaining the long delays, frustration and general "not fit for purpose" aspects of much of science publication, particularly in biological fields. An important omission however was the failure to address the credibility of what actually appears in press. There are numerous incidents of papers published in prestigious high-impact journals, later found to be wrong or deliberately fraudulent. One sad example recently was the publication of two papers from one institution in the same volume of Nature, to be followed fairly quickly by reports by other scientists of failure to reproduce the work, and then a clear admission of fraudulent aspects of the work with retraction of the papers and suicide in the lab of one of the senior mentors in the institute.*

The question arises how could *Nature* accept and publish these experiments without more evidence of their integrity. The claim of the workers was dramatic, namely that an acidic culture medium would suffice to dedifferentiate mature cells to an embryonic stem cell-like phenotype. No doubt, this was regarded as such an important breakthrough, that there was pressure on the editorial board to accept it.

From the facts detailed in the previous chapters, science and medical journals need to tighten up and reassess their reviewing processes. Perhaps when a submitted paper makes claims that might be regarded as unlikely or revolutionary, the journal editor should convene a *"special assessment committee"*, to which scientists with current expert knowledge and interest in the subject under review should be invited to contribute. Their opinions should be sought as to whether or not to proceed with processing the manuscript and who would be the most appropriate reviewers to solicit.

There might be a place for the rich journals to establish a new well-funded journal of *"Attempted Reproducibility in Science"*. Unfortunately, repeating and confirming or refuting other people's work is not regarded as productive and can be boring, so there might not be many papers submitted to this unlikely hypothetical journal. A variety of analyses would suggest that a small but significant percentage of published papers that make exciting new claims, when specifically tested, are not reproducible.

These recommendations would not be easy to follow, since most potential reviewers are very busy and are usually not paid. I suggest that they should be remunerated for undertaking a difficult and time-consuming task. Most of the high impact journals could afford to be generous and in the long run save themselves money and possibly severe embarrassment.

An alternative to an attempted reproducibility journal would be for bio-medical journals to set up a provisional pre-publication process, to safeguard the priority of the scientist as is practised by Mathematical and Physics Journals. This would allow the scientist to establish a stake for provisional priority of data that have not yet been fully verified and provide time for the work to be replicated by independent laboratories. If after a period of, say six months of open consultation on a special internet forum, the applicant would then be invited to submit the definitive article or withdraw the manuscript with dignity.

Despite the success of pre-publication in the physical sciences and mathematics, similar attempts with the biological sciences have been less successful. Websites attempting to do this have been flooded with articles, often of low quality, to an extent that cannot be controlled, and the question of credibility of the submissions has not been fully addressed.

I would suggest a possible way forwards as follows:

1. The authors would submit to a web site a pre-publication strictly limited in length and a payment to the web site of a modest handling fee of, say, $200.

2. The authors would submit six months later to the same web site evidence of credibility of their pre-publication or failure to demonstrate reproducibility, thus the pre-publication would with the passage of time have had a chance to earn a modest seal of approval. With these two steps the authors would have established the priority of their work and be in a position to submit to a journal of their choice a definitive manuscript.

3. In order to avoid the web site being swamped with "scientific junk", before the pre-publication abstract is accepted

the editors handling the web site would be free to consult any experts they wish to decide whether or not to print the pre-publication manuscript.

The above suggestions would provide anxious scientists with the opportunity to establish the priority of their research, but also would enable the editors of the journal receiving the definitive manuscript to go through the process of acceptance or rejection with at least some supporting credibility. This would have avoided the unfortunate occurrence with the acid culture report referred to above with its associated tragedy.

11.3 The Recruitment Process

Even before a scientist begins to work, there can be examples of fraud. Recently an applicant for a research post, in a laboratory in which I worked, suggested the names of suitable referees with email addresses for contact. The stupendous references raised suspicion in one member of the appointments committee. She discovered that the applicant had sent her own extra email addresses followed up by the references she had written herself. The authorities have reported that they are aware of this type of fraud and plan to clampdown fiercely on the fakers (Nature, 22 June 2017).

However, as exemplified by the Macchiarini case, scientific institutions need to do more in their scrutiny of the background and curriculum vitae of potential researchers. For a recruitment committee appointing a new staff member to work in their laboratory, the process can bear many similarities to buying a racehorse you hope will be a winner, it is always a chancy business. What is the background from which the candidate is coming? Can the details of the curriculum vitae be trusted or could the candidate be a cheat, who has forged the curriculum vitae, intercepted the contacts for the referees, so as to present fake references? Does the candidate at interview exhibit sufficient knowledge and enthusiasm for the job and the science involved? Will the candidate fit in socially and as a working colleague in the lab?

After these considerations, the appointments committee should pause to 'examine their own navel'. Does the lab flourish on friendly co-operation or does it adhere to the Machiavellian principle that fear is more effective than love, usually a mistaken philosophy? In the past, a discreet private telephone call to the candidate's previous departmental head would be the most important factor for the committee to consider, but now such an enquiry would be forbidden, as being politically unacceptable.

At the end of a two-way discussion between members of the committee and the candidate, a decision will be made and if the appointment is confirmed the lab might end up with either a Gold Cup winner, or an appointment they later regret.

It is therefore essential that science laboratories and especially hospitals and medical institutions conduct a rigorous scrutiny of all the curricula vitae of individuals applying to work in their institutions. Clear documentation must be provided and well-defined protocols must be followed meticulously. This process might be aided by modern technology which can detect and eliminate forged references.

The noble culture of science is a vocation that is often difficult to follow. These suggestions, if followed, might reduce the temptation to indulge in fake science and help to portray science as a career with high moral values.

Hope Springs Eternal in the Human Heart

(Alexander Pope, 1734).

I hope that this little book has explained the permeation of deceit throughout nature and how fake news based on lies is part of human culture including science. Efforts to be honest in science may require a brave minority swimming against the powerful stream of human behaviour and established organisations. A special case has been made for scientists to be different because the scientific method confers truth by the demonstration that *it works* and can be reproduced by other scientists independent of the original advocates. Unfortunately, scientists are no different from the rest of humanity in succumbing to the temptation to cheat, despite the certain foreknowledge that they will be found out. The unfolding of three disastrous scientific scandals, shows that unsatisfactory features persist in the management of science.

I have attempted to suggest some remedies that might help put the house in order, but it would be unrealistic to think that dishonesty in science, as in all other human pursuits can be eliminated. We can only hope that recognition of the problem can help the efforts to introduce improvements to clean up the field. For the time being, it may be prudent for science and scientists to follow these words of advice from the Buddha:

Believe nothing, even if I have said it, unless it agrees with your own reason and common sense.

Hotei God of Contentment/Happiness
© onmarkproductions.com

Fig. 11.1. Sculpture image of the Buddha (IPD).

CPSIA information can be obtained
at www.ICGtesting.com
Printed in the USA
LVHW081022131021
699877LV00005B/36